ALGEBRA PUZZLES

Build Pre-Algebra and Algebra Skills through Puzzles and Problems

Written by
Hank Garcia

Editor: Alaska Hults
Designer/Production: Moonhee Pak/Carlie Hayashi
Cover Designer: Moonhee Pak
Art Director: Tom Cochrane

Table of Contents

Introduction

Algebra and pre-algebra teach students to think about math in abstract terms. Instead of four bananas and four bananas (4 + 4), we have four bananas and an unknown quantity of bananas (4 + b).

While many students will go on to fields that do not require the daily use of algebra, they will all require the ability to think abstractly. Like any other muscle, the brain works best when it receives regular exercise. When it is asked to think in new ways, the brain builds new pathways. Pathways already built become stronger and stay active. *Algebra Puzzles* challenges students to use skills they have already acquired to solve problems that require them to think abstractly, or in ways that go beyond the typical review worksheet. The puzzles can be completed during free time over the course of the day or during a ten-minute open spot between subjects. They are a perfect sponge activity or can be assigned as "something different" for homework.

Algebra and pre-algebra vocabulary is reviewed, basic concepts are explained again, and analogies are made to familiar concepts or terms in other subject areas. *Algebra Puzzles* provides the opportunity to explore learned concepts in new ways, while challenging students to think more deeply about these concepts.

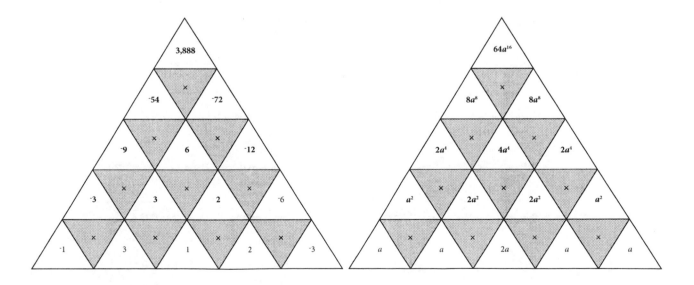

Name _____   Date _____

Integer Pyramid

Integer: a whole number that can be either greater than 0, called positive, or less than 0, called negative. Negative numbers are perfect for expressing "differences" (losses). Let's imagine that you weigh your poodle and find she is 48 pounds. You take her out running every day for a month and weigh her again and find she is 45 pounds. You could say she lost 3 pounds. You can also write *The difference in my poodle's weight between this month and last month is ⁻3 pounds.* You could draw this on a number line like this:

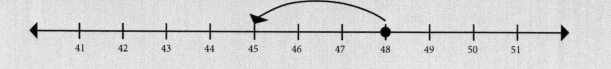

DIRECTIONS Add each pair of adjoining integers. Write the sum in the white triangle centered above each pair.

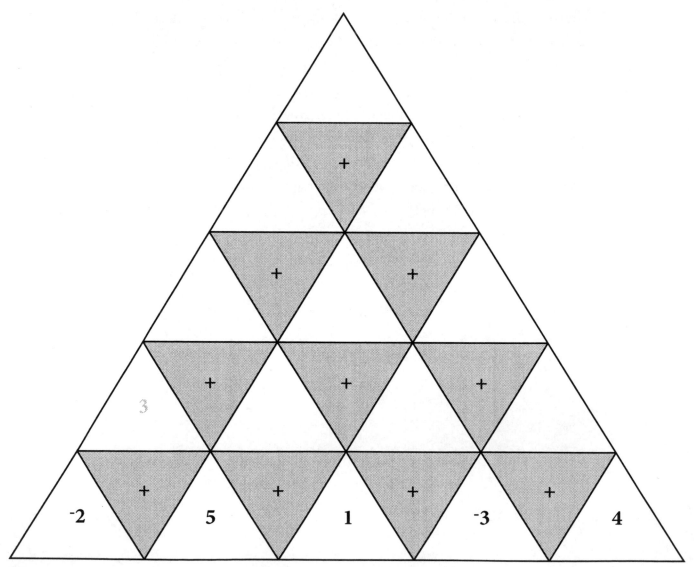

Multiplication Pyramid

If both integers are positive or negative, then the product is positive. If one integer is positive and the other is negative, then the product is negative.

 Multiply each pair of adjoining integers. Write the product in the white triangle centered above each pair.

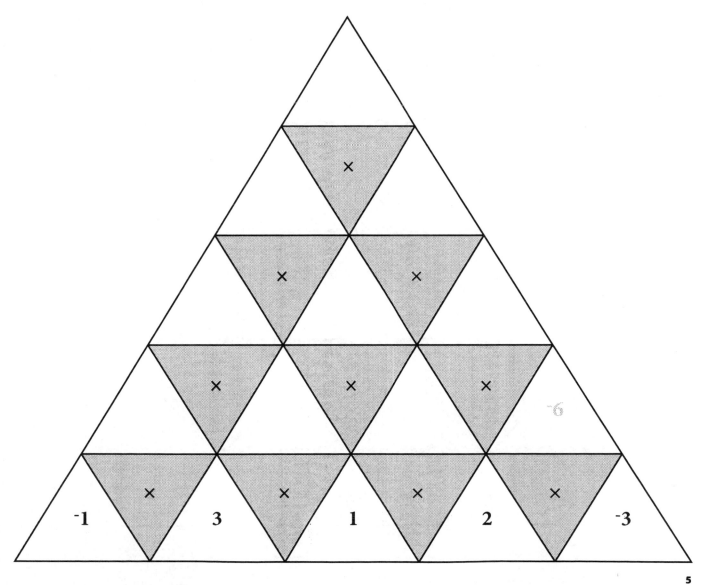

Mixed Operations Pyramids

 **DIRECTIONS** Add or multiply each pair of adjoining integers as indicated. Write the answer in the white triangle centered above each pair.

1

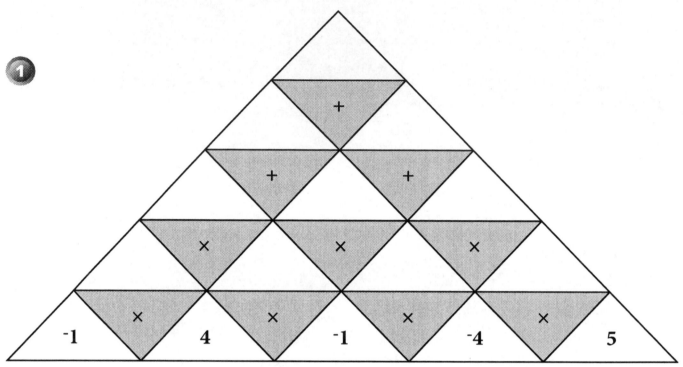

2

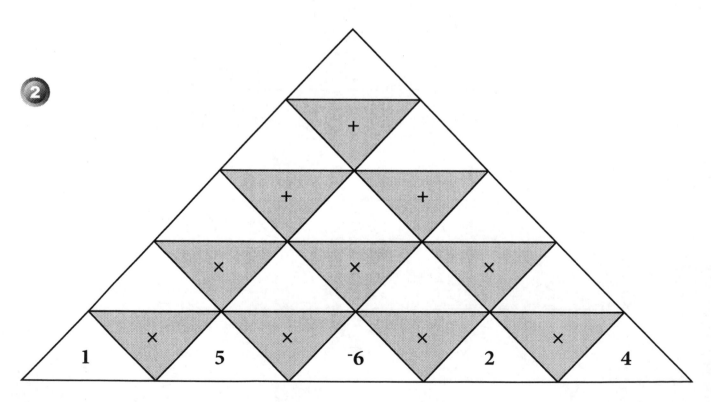

Integer Wheel

 **REMEMBER!** If both integers are positive or negative, then the product is positive. If one integer is positive and the other is negative, then the product is negative.

 **DIRECTIONS** The product of any 3 factors in a straight line across the wheel should be the same. Use the integers in the box next to each wheel to complete each 3-factor spoke. Write the product of each wheel's spokes.

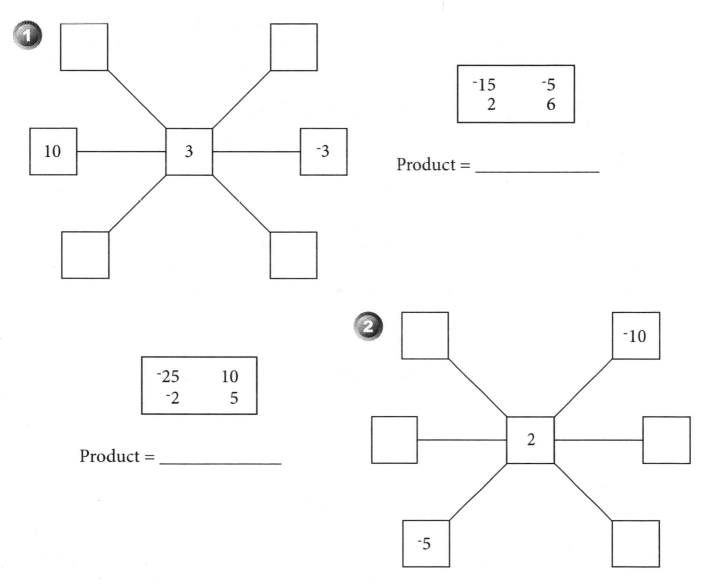

1

10 — 3 — -3

| -15 | -5 |
| 2 | 6 |

Product = _____

| -25 | 10 |
| -2 | 5 |

Product = _____

2

-10

2

-5

3 If one number in the wheel is zero, what must the product of every "spoke" be?

_____

Name _____   Date _____

It's Magic!

A magic square is a grid in which each row, each column, and the two corner-to-corner diagonals all have the same sum. This magic square has a magic number of 12.

7	0	5
2	4	6
3	8	1

DIRECTIONS Use the information given to determine the magic number of each square. Then write numbers in the boxes to complete the squares. There is more than one solution to some of the squares, so be prepared to share your answer with the class.

 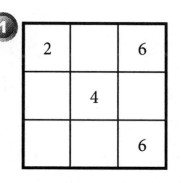

1

2		6
	4	
		6

2

14		8
-1	5	
		-4

3

		6
2	2	
-2		-2

Look for a pattern in the completed boxes. Examine the relationship between the number in the center box and the magic number.

4 What is the magic number of a magic square in which the center number is 8?

_____

DIRECTIONS Complete the remaining magic squares.

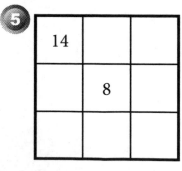 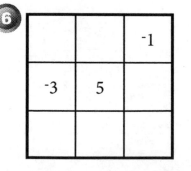 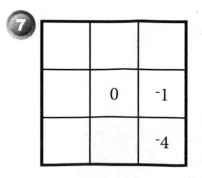

5

14		
	8	

6

		-1
-3	5	

7

	0	-1
		-4

"Sum" More Magic

 **DIRECTIONS** Add the given number to each number in the magic square. Three numbers have been completed for you. Look at the patterns. Then answer the question below.

1

8	0	7
4	5	6
3	10	2

+ 3   =

		10
	8	
		5

2

-1	4	6
10	3	-4
0	2	7

+ 5   =

3

4	-9	5
1	0	-1
-5	9	-4

+ 6   =

4 Does the new square follow the rules of a magic square? _____

Name _____   Date _____

Multiplication Mix-Up

Etymology is the study of the history of words. For example, **malaria** (a deadly disease) comes from the words *mal* and *aria,* which are Italian words meaning "bad" and "air." These words reflect the ancient belief that a person caught malaria by breathing bad air. (We know now that mosquitoes are to blame.)

The original meaning of the word **product** is "a mathematical quantity obtained by means of multiplication" and it originated in the early 1400s from the Latin word *producere* (to bring forth).

 Multiply each pair of factors and record the product in the table. Notice that the factors are out of order (when compared to a typical multiplication table).

×	5	⁻3	0	8	4	2	⁻9
9							
1							
6		⁻18					
5						10	
⁻4							
7							
⁻3							

Variables

Pronouns were added to languages so early on that their exact appearance in the development of language is unknown. Maybe pronouns like **I** were invented before first names. We really do not know. But we do know that words like **I**, **me**, **mine**, and **it** make communication easier.

Variables are the "pronouns" of math. Instead of saying "unknown quantity," we say *x*. Instead of saying "that other unknown quantity," we say *y*. Let's imagine that we know we need to buy two apples for each person coming to our party, but we do not know how many people are coming to the party. We can write 2*x* to represent "twice the unknown number of people coming to the party."

 Multiply each pair of factors and record the product in the table. Notice that some factors are variables. The examples show how to record a product that includes a variable.

×	5	⁻1	*n*	12	⁻*y*	2	0
z	5z		zn				
⁻6							
4							
9							
7							
⁻3							
x							

Name _____   Date _____

More Multiplication Mix-Up

Multiplying a factor by itself is also known as **squaring** a number. If you square a variable, you write it by adding the exponent at the top right of the letter.

$$x \cdot x = x^2$$

 Multiply each pair of factors and record the product in the table. Notice that some factors are variables. The examples show how to record a product that includes a variable.

×	11	‾3	8	‾9	s	2	18
‾7	‾77						
1							
‾5							
8					8s		
‾4							
‾s							
15							

Name _____ Date _____

Addition Map

Adding a negative is like subtracting a positive.

$$2 + (-1) = 1 \qquad 2 - 1 = 1$$

 **DIRECTIONS** Start anywhere along an edge. Make a "trail" by adding together a set of integers. Each trail must end up at and equal the number in the final box (5). There is more than one correct path. Find as many trails as you can.

⁻6	⁻4	6	⁻5	⁻2
5	3	2	⁻1	8
⁻2	7	1	⁻2	3
⁻9	⁻1	2	1	4
	5			

Additional Maps

 **DIRECTIONS** Start anywhere along an edge. Make a "trail" by adding together a set of integers. Each trail must end up at and equal the number in a final box (⁻7 or ⁻3). There is more than one correct path. Find as many trails as you can.

4	3	1	⁻1	9
2	⁻8	⁻1	2	5
⁻2	3	⁻2	1	⁻7
		⁻7		

8	7	1	⁻6	⁻5
⁻2	3	1	⁻1	⁻3
6	⁻1	2	4	2	5
	⁻7	3	⁻2	1	⁻4
		⁻3		

Algebra Puzzles © 2006 Creative Teaching Press

Name _____   Date _____

Check It Out!

DIRECTIONS Put a checkmark in the box if the category on the left fits the term.

	$2x^2$	$3x$	$12x^2$	4	x
Positive Number					
Quadratic Term	✓				
Composite					
Factor of $24x^2$					
Reciprocal of $\frac{1}{3}x$					
Rational Number					
A Constant					

Name _____   Date _____

Check, Check

 **DIRECTIONS** Put a checkmark in the box if the category on the left fits the term.

	7	$a + 2$	0	11	$2b^2$
Integer	✓				
Polynomial					
Variable Expression					
Equal to Its Absolute Value					
Exponential Form					
Monomial					
Prime					

Nice Cube

Cubed: multiplying a number by itself, and then multiplying that product by the original number.

$$x \cdot x \cdot x = x^3$$

A **cube** is a 3-dimensional solid that has the same length, width, and height. To find its **volume**, or the amount of space it takes up, you would multiply the length times the width times the height. For example, the calculation to find the volume of a cube with edges that are 2 units long would look like the following:

$$2 \times 2 \times 2 = 8$$

 **DIRECTIONS** Use a ruler and pencil to show each exponential expression on the cube as in the example above.

1 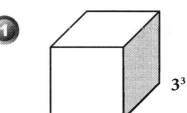 3^3

3 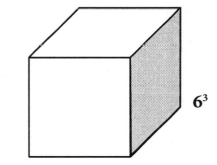 6^3

2 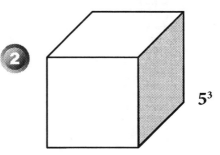 5^3

4 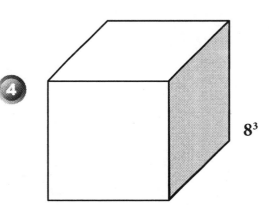 8^3

Add or Subtract?

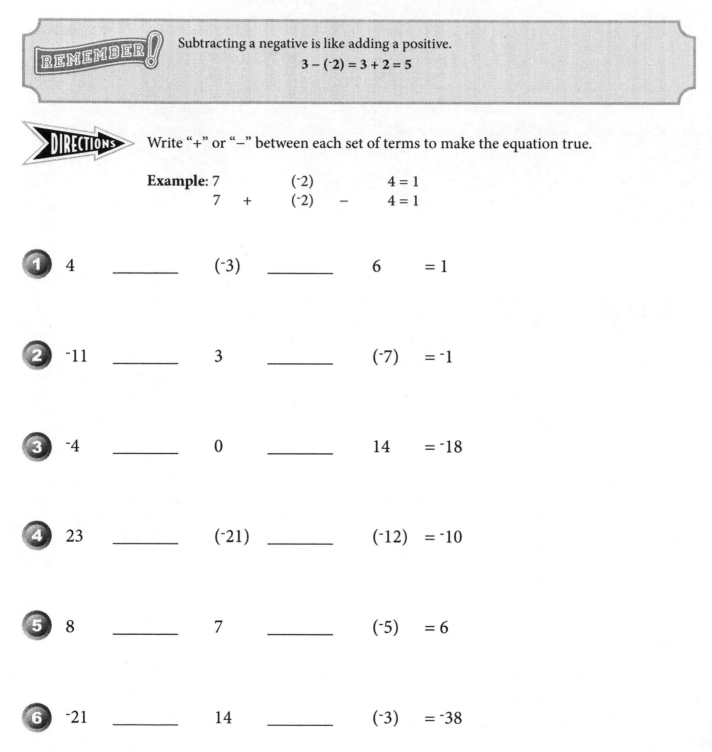

REMEMBER! Subtracting a negative is like adding a positive.

3 − (⁻2) = 3 + 2 = 5

DIRECTIONS   Write "+" or "−" between each set of terms to make the equation true.

Example: 7   (⁻2)   4 = 1

7   +   (⁻2)   −   4 = 1

1 4 _____ (⁻3) _____ 6   = 1

2 ⁻11 _____ 3 _____ (⁻7)   = ⁻1

3 ⁻4 _____ 0 _____ 14   = ⁻18

4 23 _____ (⁻21) _____ (⁻12)   = ⁻10

5 8 _____ 7 _____ (⁻5)   = 6

6 ⁻21 _____ 14 _____ (⁻3)   = ⁻38

Name _____ Date _____

Add or Subtract with Variables

 Write "+" or "−" between each set of terms to make the equation true.

Example: 5   (^-x) $3 = 2 - x$
             5   +   (^-x)   − $3 = 2 - x$

1 6 _____ 2 _____ 1 $= 5$

2 $^-5$ _____ ($^-3$) _____ n $= n - 8$

3 b _____ 9 _____ ($^-8$) $= b - 17$

4 7 _____ x _____ ($^-2$) $= 9 + x$

5 3 _____ 7 _____ ($^-5$) $= ^-9$

6 $^-11$ _____ 14 _____ ($^-3$) $= ^-22$

Name _____   Date _____

Like Terms

Like Terms: quantities that have the same variables.

$$2n, n, 4n \quad \text{but not} \quad 2n^2$$

You can only add or subtract like terms.

$$2n^4 - n^4 + 3n - 2n = n^4 + n$$

DIRECTIONS   Write "+" or "−" between each set of terms to make the equation true.

1 $^-4$ _____ 5 _____ 8 $= {}^-1$

2 $2n$ _____ 2 _____ n $= n + 2$

3 $^-n^2$ _____ $5n$ _____ $2n^2$ $= 5n + n^2$

4 $2n^3$ _____ n _____ n^3 $= n^3 - n$

5 $3n^2$ _____ $4n^2$ _____ n^3 $= n^3 - n^2$

6 ^-n _____ $3n$ _____ $(^-3n^2) = 3n^2 - 4n$

Name _____   Date _____

Factor Fun

Factor: a number or variable that is multiplied to make a product.

 Circle any four factors in a row that result in the product **24n**. Answers can be horizontal, vertical, or diagonal. See the example below.

	A	B	C	D	E	F	G	H	I	J
1	1	4	8	$2n$	3	4	n	2	n	8
2	4	6	12	1	2	1	2	1	1	n
3	n	n	$6n$	1	4	6	n	2	n	3
4	4	1	3	$2n$	6	2	1	3	6	1
5	$3n$	4	n	1	2	$6n$	2	$3n$	2	$8n$
6	$2n$	4	2	12	3	n	12	1	$2n$	6

Name _____   Date _____

More Factor Fun

 Circle any four factors in a row that result in the product **16a**. Answers can be horizontal, vertical, or diagonal. See the example below.

	A	B	C	D	E	F	G	H	I	J
1	1	4	8	2a	3	4	a	2	⁻a	8
2	4	⁻1	12	1	2	1	⁻2	⁻2	1	a
3	a	⁻1	2a	1	4	2	8	1	⁻a	⁻2
4	4	1	2	⁻8	6	2	⁻1	a	1	⁻1
5	⁻a	4	a	1	2	4a	2	2a	2	⁻a
6	2a	⁻4	2	⁻1	1	a	⁻a	1	⁻4	1

Do-It-Yourself Equations

DIRECTIONS Circle any three or four terms or factors that will make an equation horizontally, vertically, or diagonally. Use any operation as long as the equation is true. See the example below.

	A	B	C	D	E	F	G	H	I	J
1	a	1	2	$2a$	-2	$4a$	2	-4	-3	6
2	$3a$	2	-1	a	1	$a \times 2 \times 1 = 2a$				$6a$
3	8	2	$2a$	9	2	4	$3a$	-a	4	6
4	$2a$	10	-a	5	$4a$	2	3	-5	a	2
5	-a	1	-1	a	-2	-6	12	4	-4	$2a$
6	$3a$	1	2	4	2	-1	a	1	2	-1
7	a	$8a$	2	$2a$	2	-1	-8	1	0	-1
8	$2a$	1	2	a	3	-5	a	2	4	-2

More Do-It-Yourself Equations

 **Circle any three or four terms or factors that will make an equation horizontally, vertically, or diagonally. Use any operation as long as the equation is true. See the example below.**

	A	B	C	D	E	F	G	H	I	J
1	⁻5	1	x	2	3	4	x $+$ $6x$ $=$ $7x$			8
2	⁻2	⁻6	12	4	⁻4	1	⁻x	1	⁻1	x
3	10	⁻x	5	$4x$	2	$2x$	9	2	4	3
4	3	1	x	4	$6x$	2	x	3	6	3
5	2	4	x	1	2	$2x$	⁻2	$4x$	2	$2x$
6	$3x$	2	1	x	1	4	3	$4x$	$2x$	$6x$

Magic or Math?

Memorize the following directions and you can appear to your friends and family to have mastered a bit of magic. But the "trick" isn't magic at all—it's algebra!

Follow the prompts to fill in the missing number. Compare your responses to the algebraic equation next to your answer. Then ask a classmate to complete each step and fill in his or her response.

DIRECTIONS

	Direction	Algebra	Your Answer	Your Friend's Answer
1	Pick a number greater than zero.	n	___	___
2	Add the number that follows it.	$n + (n + 1) = 2n + 1$	___ + ___ = ___	___ + ___ = ___
3	Add 9 to your total.	$2n + 1 + 9 = 2n + 10$	___ + 9 = ___	___ + 9 = ___
4	Divide your total by 2.	$(2n + 10) \div 2 = n + 5$	___ ÷ 2 = ___	___ ÷ 2 = ___
5	Subtract the original number.	$n + 5 - n = 5$	___ − ___ = 5	___ − ___ = 5

The answer is always 5.

More Magic or Math?

 **DIRECTIONS** Follow the prompts to fill in the missing number. Compare your responses to the algebraic equation next to your answer. Then ask a classmate to complete each step and fill in his or her response.

Direction	Algebra	Your Answer	Your Friend's Answer
1 Think of the number of the month in which you were born.	m	_____	_____
2 Multiply by 4.	$4m$	$4 \times$ _____ = _____	$4 \times$ _____ = _____
3 Add 16.	$4m + 16$	_____ + 16 = _____	_____ + 16 = _____
4 Multiply by 5.	$5(4m + 16) = 20m + 80$	_____ × 5 = _____	_____ × 5 = _____
5 Subtract 7.	$20m + 80 - 7 = 20m + 73$	_____ – 7 = _____	_____ – 7 = _____
6 Multiply by 5.	$5(20m + 73) = 100m + 365$	_____ × 5 = _____	_____ × 5 = _____
7 Add the day of the month of your birthday.	$100m + 365 + d$	_____ + _____ = _____	_____ + _____ = _____
8 Subtract 365.	$100m + 365 + d - 365 =$ $100m + d$	_____ – _____ = _____	_____ – _____ = _____

The hundreds and thousands place (if there is one) represent the number of the month. The tens and ones place are the date. For example, December 18 = 1218 and May 5 = 505.

Try This One!

Follow the prompts to fill in the missing number. Compare your responses to the algebraic equation next to your answer. Then ask a classmate to complete each step and fill in his or her response.

▼ **DIRECTIONS**

	Direction	Algebra	Your Answer	Your Friend's Answer
1	Think of a two-digit number (positive).	n	_____	_____
2	Add 25.	$n + 25$	_____ + 25 = _____	_____ + 25 = _____
3	Multiply by 2.	$2(n + 25) = 2n + 50$	_____ × 2 = _____	_____ × 2 = _____
4	Subtract 4.	$2n + 50 - 4 = 2n + 46$	_____ − 4 = _____	_____ − 4 = _____
5	Divide by 2.	$(2n + 46) \div 2 = n + 23$	_____ ÷ 2 = _____	_____ ÷ 2 = _____
6	Subtract the original number.	$n + 23 - n = 23$	_____ − _____ = _____	_____ − _____ = _____

The answer is always 23!

Property Matching Game

Commutative $a + b = b + a$ **or** $a \times b = b \times a$
Associative $a + (b + c) = (a + b) + c$
Distributive $a(b + c) = ab + ac$

DIRECTIONS Match each equation on the right with the property it illustrates on the left.

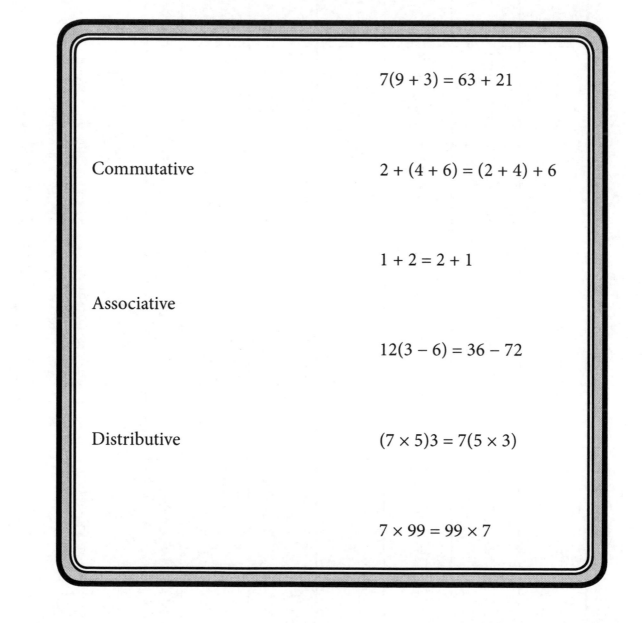

$7(9 + 3) = 63 + 21$

Commutative

$2 + (4 + 6) = (2 + 4) + 6$

$1 + 2 = 2 + 1$

Associative

$12(3 - 6) = 36 - 72$

Distributive

$(7 \times 5)3 = 7(5 \times 3)$

$7 \times 99 = 99 \times 7$

More Property Matching

Identity Property of Addition $a + 0 = a$

Identity Property of Multiplication $a \times 1 = a$

Transitive Property   If $a = b$ and $b = c$, then $a = c$

DIRECTIONS Match each equation on the right with the property it illustrates on the left.

Identity Property of Multiplication

$(3 + n) + 2 = 3 + (n + 2)$

Identity Property of Addition

$12n \times 1 = 12n$

Transitive

If $x + 4 = 5$ and $5 = 1 + 4$, then $x + 4 = 1 + 4$

Commutative

$5(n - 3) = 5n - 15$

Associative

$15,678,237 + (n \times 0) = 15,678,237$

Distributive

$43 \times 3 = 3 \times 43$

Name _____   Date _____

Addition Mix-Up 1

A **term** is a number, or the product of a number and one or more variables. Some examples are **4**, **2x**, and **3x + y**.

 **DIRECTIONS**   Add each pair of terms and record the sum in the table.

+	n	4	2	-3	8	6	-1
7							
-2							
9							
1	n + 1						
3							
m							
-5			-3				

Name _____   Date _____

Addition Mix-Up 2

 **DIRECTIONS**   Add each pair of terms and record the sum in the table.

+	*a*	1	2	13	⁻6	5	⁻8
12							
⁻9							
⁻1							
14	*a* + 14						
3							
7							
⁻*b*				13 − *b*			

Name _____   Date _____

Addition Mix-Up 3

 **DIRECTIONS**   Add each pair of terms and record the sum in the table.

+	-4	12	2	15	y	6	-1
-3							
-7							
8							
z	$z - 4$						
13							
9							
11							

Balancing Act

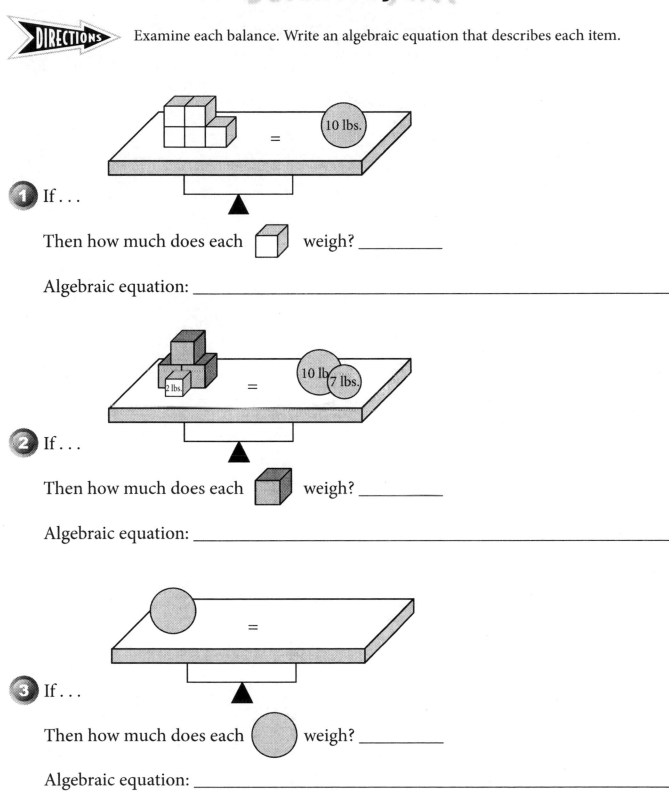

DIRECTIONS   Examine each balance. Write an algebraic equation that describes each item.

1 If . . .

Then how much does each ☐ weigh? _____

Algebraic equation: _____

2 If . . .

Then how much does each ☐ weigh? _____

Algebraic equation: _____

3 If . . .

Then how much does each ◯ weigh? _____

Algebraic equation: _____

Name _____ Date _____

More Balancing Act

 Examine the illustrations and then circle the side of the balance that best answers each question.

If . . .

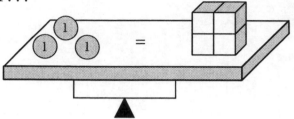

and

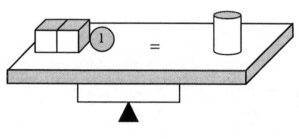

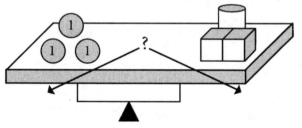

1 Which way will the balance tilt?

_____

_____

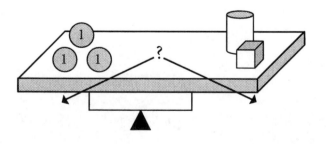

2 Which way will the balance tilt?

_____

_____

Algebra Puzzles © 2006 Creative Teaching Press

Name _____ Date _____

Balanced Logic

Algebra Puzzles © 2006 Creative Teaching Press

 **DIRECTIONS** Find the value of *b*.

If . . .

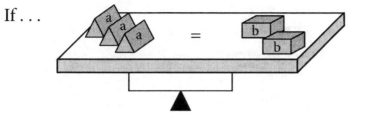

and

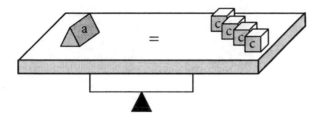

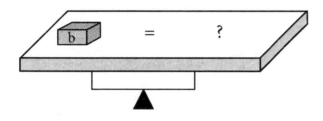

1 What does *b* equal relative to *c*? _____

2 Explain in words or equations how you figured this out.

_____

_____

_____

_____

Positives and Negatives

Here's a fun way to remember the rule about multiplying or dividing positive and negative integers. Think of one of the other students in your school and how well you get along.

- If someone likes you (+) and you like him or her (+), that's positive (+).
- If someone likes you (+), but you don't really like him or her (−), that's no fun (−).
- If someone doesn't like you (−), but you like him or her (+), that's no fun either (−).
- If someone doesn't like you (−) and you don't like him or her either (−), that's just fine (+) because you probably won't spend much time around each other!

DIRECTIONS   Write "×" or "÷" between each set of terms to make the equation true.

Example: 16   2   4 = 32
             16   ÷   2   ×   4 = 32

1   12   _____   3   _____   6   = 6

2   6   _____   2   _____   ⁻3   = ⁻4

3   ⁻5   _____   ⁻3   _____   2   = 30

4   24   _____   ⁻8   _____   7   = ⁻21

5   36   _____   9   _____   ⁻5   = ⁻20

6   ⁻21   _____   7   _____   ⁻7   = 21

Multiply or Divide?

DIRECTIONS Write "×" or "÷" between each set of terms to make the equation true.

Example: 16   2   4 = 32

16 ÷ 2 × 4 = 32

1 ⁻7 _____ 4 _____ 2 = ⁻14

2 10 _____ ⁻2 _____ ⁻8 = 40

3 15 _____ 2 _____ ⁻2 = ⁻60

4 4 _____ ⁻8 _____ 8 = ⁻4

5 12 _____ 2 _____ ⁻6 = ⁻1

6 ⁻9 _____ ⁻3 _____ ⁻3 = ⁻9

Algebra Puzzles © 2006 Creative Teaching Press

Divide or Multiply?

Write "×" or "÷" between each set of terms to make the equation true.

Example: 16   2   4 = 32

16   ÷   2   ×   4 = 32

1   16 _____ ⁻4 _____ ⁻2   = 2

2   8 _____ 5 _____ ⁻6   = ⁻240

3   ⁻2 _____ ⁻2 _____ 2   = 2

4   12 _____ ⁻3 _____ 6   = ⁻24

5   20 _____ 2 _____ ⁻10   = ⁻100

6   ⁻42 _____ 7 _____ ⁻3   = 18

Exponents

Exponent: a number that indicates how many times the base is used as a factor.

$$\text{base} \rightarrow 3^{2} \xleftarrow{\text{exponent}}$$

$$= 3 \times 3$$

Bases with the same number or variable can be multiplied or divided. The base stays the same, but when you multiply the terms, you add the exponents.

$$a^5 \times a^2 = a^7$$
$$4a^5 \times 2a^2 = 8a^7$$

When you divide the terms, you subtract the exponents.

$$10n^5 \div 2n^3 = 5n^2$$

> **DIRECTIONS** Write "×" or "÷" between each set of terms to make the equation true.

1 $2n^2$ _____ 2 _____ n^2 = $4n^4$

2 $2a^3$ _____ $5b$ _____ $^-b^3$ = $^-10a^3b^4$

3 $^-n^2$ _____ ^-n _____ $^-n^3$ = $^-n^6$

4 $8x^4$ _____ $2x$ _____ $^-x^3$ = $^-4$

5 $20n$ _____ n^3 _____ $10n^2$ = $2n^2$

6 $30a^{10}$ _____ $6a^4$ _____ a^5 = $5a$

GEMS

In a math problem with more than one operation, you must follow the correct order of operations to arrive at the correct answer.

GEMS
First   **Grouping symbols**
Then   **Exponents**
Then   **Multiplication or division**
Then   **Subtraction or addition**

DIRECTIONS   Use the space below and the correct order of operations to solve the problem.

$$4 + 6 \times 3 - 4 \div 1 + 2 \times 3 - (7 + 2) - 2 \times 3 - 9 \div 3 + 2 = \underline{\hspace{2cm}}$$

PEMDAS

Here is another acronym you can use to remember the correct order of operations.

PEMDAS

First	**Parentheses**
Then	**Exponents**
Then	**Multiplication**
or	**Division**
Then	**Addition**
or	**Subtraction**

 Use the space below and the correct order of operations to solve the problem.

$(9 - 2) \times (4 + 3) + 3 - 12 \div 4 \div 3 + 16 - 1 + (7 + 2) \times 5 - 16 \div 2 =$ _____

_____

_____

_____

_____

_____

_____

_____

_____

_____

Name _____   Date _____

One Powerful Problem

An exponential expression has a base and an exponent.

A base can be an integer, a fraction, a decimal, a variable, or a combination of any of those inside parentheses. You simplify the expression as much as possible, then solve it as usual.

$$(4 \times a)^2$$
$$(4a)^2$$
$$(4a)(4a)^2$$
$$16a^2$$

DIRECTIONS Use the space below and the correct order of operations to solve the problem.

$$(2 + 2) \times 3 + 4^2 + 1 + 8 \div 4 - 5 - (1 + 2)^2 + 3 - 12 \div 3 + 2^2 = _____$$

_____

_____

_____

_____

_____

_____

_____

_____

Grouping Symbols

Anything inside of a grouping symbol is simplified first in a math problem with more than one operation.
Here are three common kinds of grouping symbols:

Brackets   [ ]

Parentheses   ( )

Fraction Bar   — or /

If there are grouping symbols *inside of* grouping symbols, work inside to outside.

$$[(3 + 4)^2 + (2 + 5)^2] + 2$$
$$[7^2 + 7^2] + 2$$
$$[49 + 49] + 2$$
$$98 + 2$$
$$= 100$$

▷DIRECTIONS▷   Use the space below and the correct order of operations to solve the problem.

$$[9 - 7 + (18 \div 3)] \div 4 + (14 \div 7) + 3 - 11 + [(8 - 6) \times 5] - 20 \div 5 = \underline{\quad\quad}$$

_____

_____

_____

_____

_____

_____

_____

_____

What Is "Radical 4"?

First Line	Second Line
Start	**Start**
1. (⁻4, 0)	1. (1, 2)
2. (⁻3, 0)	2. (1, 0)
3. (⁻2, ⁻3)	3. (4, 0)
4. (0, 3)	4. (3, 0)
5. (5, 3)	5. (3, 2)
Stop	6. (3, ⁻3)
	Stop

DIRECTIONS Plot the coordinates to find the answer to the question.

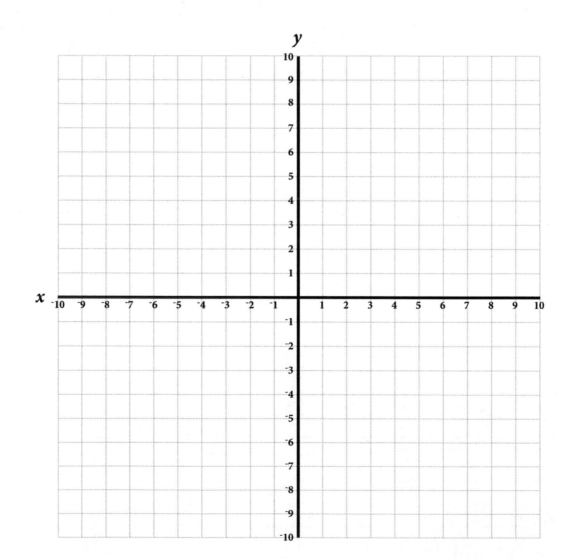

Name _____ Date _____

Picture This

Which operations are commutative and associative?

First Line		Second Line		Third Line	
Start		**Start**		**Start**	
1. (3, ⁻3)	8. (6, ⁻8)	1. (3, ⁻2)	8. (⁻2, ⁻1)	1. (⁻6, 8)	8. (⁻6, 3)
2. (4, ⁻3)	9. (5, ⁻6)	2. (2, ⁻2)	9. (⁻1, ⁻2)	2. (⁻5, 8)	9. (⁻6, 5)
3. (5, ⁻5)	10. (4, ⁻8)	3. (⁻1, 3)	10. (1, ⁻2)	3. (⁻5, 6)	10. (⁻8, 5)
4. (6, ⁻3)	11. (3, ⁻8)	4. (⁻1, 5)	11. (3, 0)	4. (⁻3, 6)	11. (⁻8, 6)
5. (7, ⁻3)	12. (4, ⁻5)	5. (1, 5)	12. (4, 0)	5. (⁻3, 5)	12. (⁻6, 6)
6. (6, ⁻5)	13. (3, ⁻3)	6. (1, 3)	**Stop**	6. (⁻5, 5)	13. (⁻6, 8)
7. (7, ⁻8)	**Stop**	7. (⁻2, 0)		7. (⁻5, 3)	**Stop**

DIRECTIONS   Plot the coordinates to find the answer to the question.

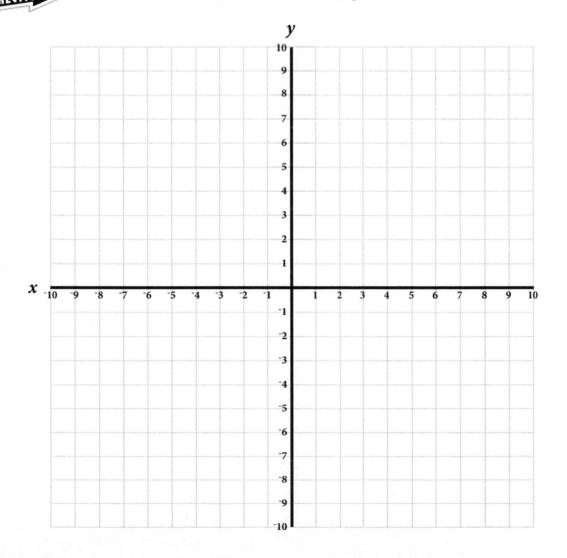

Algebra Puzzles © 2006 Creative Teaching Press

More Picture This

Which country first used coordinates to survey land?

First Line	Second Line	Third Line		Fourth Line		Fifth Line		Sixth Line	
Start	**Start**	**Start**		**Start**		**Start**		**Start**	
1. (3, ⁻1)	1. (4, ⁻2)	1. (⁻6, 5)	8. (⁻4, 1)	1. (5, ⁻6)	6. (8, ⁻10)	1. (⁻1, 2)	7. (2, 0)	1. (⁻10, 8)	7. (⁻7, 5)
2. (6, ⁻1)	2. (5, ⁻2)	2. (⁻2, 5)	9. (⁻4, 2)	2. (5, ⁻5)	7. (7, ⁻10)	2. (0, 2)	8. (2, ⁻3)	2. (⁻7, 8)	8. (⁻9, 5)
3. (6, ⁻4)	3. (5, ⁻3)	3. (⁻2, 3)	10. (⁻2, 2)	3. (10, ⁻5)	8. (7, ⁻6)	3. (1, 1)	9. (1, ⁻3)	3. (⁻7, 7)	9. (⁻9, 4)
4. (4, ⁻4)	4. (4, ⁻3)	4. (⁻3, 3)	11. (⁻2, 0)	4. (10, ⁻6)	9. (5, ⁻6)	4. (2, 1)	10. (1, 0)	4. (⁻9, 7)	10. (⁻7, 4)
5. (4, ⁻6)	5. (4, ⁻2)	5. (⁻3, 4)	12. (⁻6, 0)	5. (8, ⁻6)	10. (5,⁻5)	5. (3, 2)	11. (⁻1, 2)	5. (⁻9, 6)	11. (⁻7, 3)
6. (3, ⁻6)	**Stop**	6. (⁻5, 4)	13. (⁻6, 5)		**Stop**	6. (4, 2)	**Stop**	6. (⁻7, 6)	12. (⁻10, 3)
7. (3, ⁻1)		7. (⁻5, 1)	**Stop**						13. (⁻10, 8)
Stop									**Stop**

> **DIRECTIONS** Plot the coordinates to find the answer to the question.

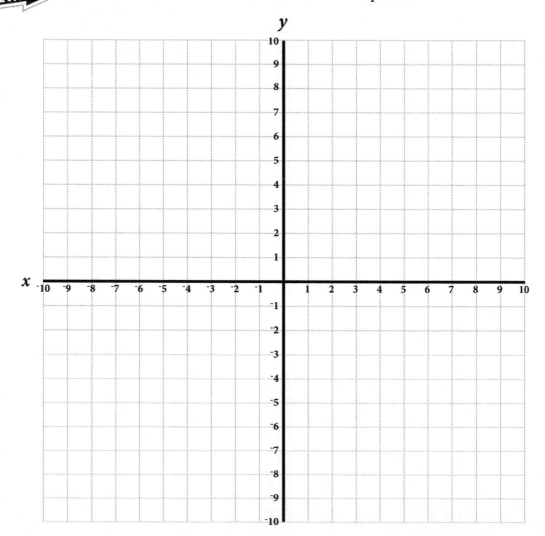

Algebra Puzzles © 2006 Creative Teaching Press

Fantastic 4

DIRECTIONS ➤ Use four 4s and any operation sign to make an equation that equals the number of the problem. See problems 1 and 7 for examples.

1 $\frac{4}{4} \times \frac{4}{4} = 1$

2 _____

3 _____

4 _____

5 _____

6 _____

7 $\frac{44}{4} - 4 = 7$

8 _____

9 _____

10 _____

Name _____   Date _____

Fabulous 5

 **DIRECTIONS**   Use five 5s and any operation sign to make an equation that equals the number of the problem. See problems 1 and 6 for examples.

1 $^{55}/_5 - 5 - 5 = 1$

2 _____

3 _____

4 _____

5 _____

6 $^5/_{0.5} - 5 + {}^5/_5 = 6$

7 _____

8 _____

9 _____

10 _____

Algebra Puzzles © 2006 Creative Teaching Press

Name _____   Date _____

Spectacular 6

DIRECTIONS Use six 6s and any operation sign to make an equation that equals the number of the problem. See problem 1 for an example.

1 $\frac{66}{66} + 6 - 6 = 1$

2 _____

3 _____

4 _____

5 _____

6 _____

7 _____

8 _____

9 _____

10 _____

Sensational 7

Absolute Value is the distance of an integer from zero. 7 and −7 are both 7 ones from zero, so they both have the absolute value of 7. We write absolute value like so:

$$|-7| \text{ and } |7|$$

 **DIRECTIONS**   Use seven 7s and any operation sign to make an equation that equals the number of the problem. See problems 2 and 8 for examples.

1 _____

2 $(\frac{7}{0.7} - |7| - \frac{7}{7}) \times \frac{7}{7} = 2$

3 _____

4 _____

5 _____

6 _____

7 _____

8 $(7 \times 7 + \frac{7}{7}) - (7 \times 7) + 7 = 8$

9 _____

10 _____

Name _____   Date _____

Impossible

 **DIRECTIONS** Fill in the blanks with various operations. Find the one answer that is <u>not</u> possible. Then unscramble the letters of your answers and write them on the blanks below to find what your parents might do if you mastered algebra this year.

Example: 3 _____ 5 _____ 2 =   1) 13   2) 10   3) 0   4) ⁻10
  $3 \times 5 - 2 = 13$ (so #1 can be correct)
  $3 + 5 + 2 = 10$ (so #2 can be correct)
  $3 - 5 + 2 = 0$ (so #3 can be correct)
but there is no way to make ⁻10 with a positive 3, 5, and 2. The answer is #4.

1 4 _____ 2 _____ 6 =   **a)** 7   **b)** 12   **c)** 0   **d)** 8

2 _____ 2 _____ 3 =   **e)** 5   **f)** 12   **g)** 7   **h)** 2

3 ⁻5 _____ ⁻6 _____ 3 =   **i)** ⁻27   **j)** ⁻8   **k)** ⁻14   **l)** 4

4 ⁻1 _____ 0 _____ 1 =   **m)** 0   **n)** ⁻1   **o)** ⁻2   **p)** 2

5 6 _____ n _____ 2 =   **q)** $12n$ **r)** $3n$ **s)** $8n$ **t)** $6n - 2$

_____ __r__ _____ _____ _____ _____

Name _____   Date _____

Not That One

 **DIRECTIONS** Find the one answer that is <u>not</u> an expression or description of the first term. Then unscramble the letters of your answers and write them on the blanks below to find what you should have to achieve success in life.

1 $5^2 =$

    a) 5×2 **c)** 25

    b) "5 squared"   **d)** 5×5

2 $n^3 =$

    e) $n \times n \times n$ **g)** $n \times 3$

    f) "n cubed"   **h)** "n to the power of 3"

3 slope =

    i) rise/run   **k)** $y - y/x - x$

    j) incline of a line   **l)** x-axis

4 $x^2 =$

    m) $x \times x$ **o)** $2x$

    n) $(x)(x)$ **p)** quadratic term

5 $|{}^-7| =$

    q) 7   **s)** ${}^-7$

    r) $-({}^-7)$ **t)** 7 spaces from origin

_____ _____ _____ _____ _____

Mystery Pyramid

Add the variables in each pair of small triangles to find the sum in the triangle above each pair. Use the known values to find the missing values.

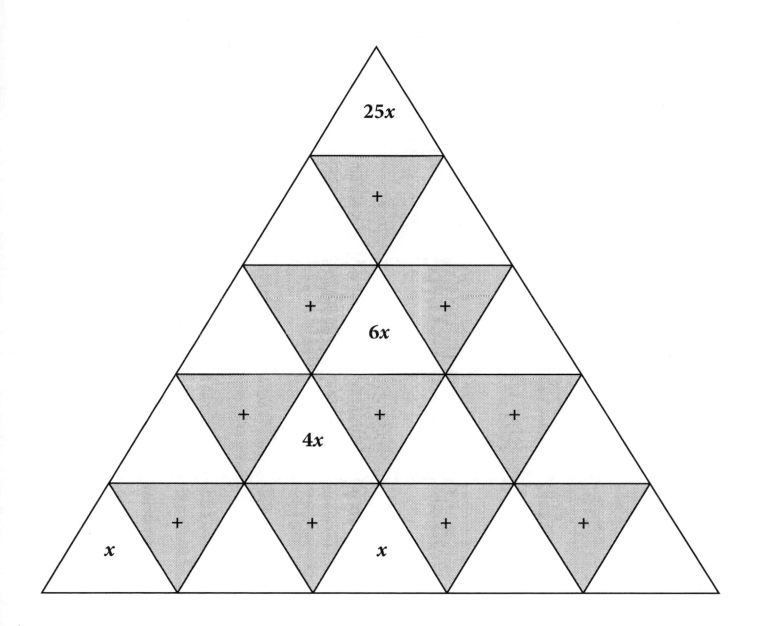

Multiplication Mystery Pyramid

 **DIRECTIONS** Multiply the variables in each pair of small triangles to find the product in the triangle above each pair. Use the known values to find the missing values.

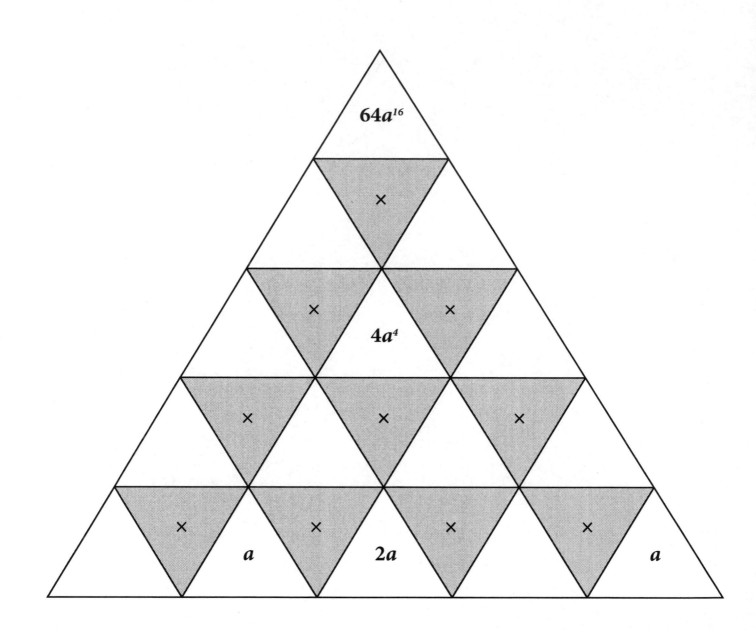

$64a^{16}$

$4a^4$

a $2a$ a

Simple Fractions

 Add the fractions in each pair of small triangles to find the final sum. Keep all
fractions in simplest form.

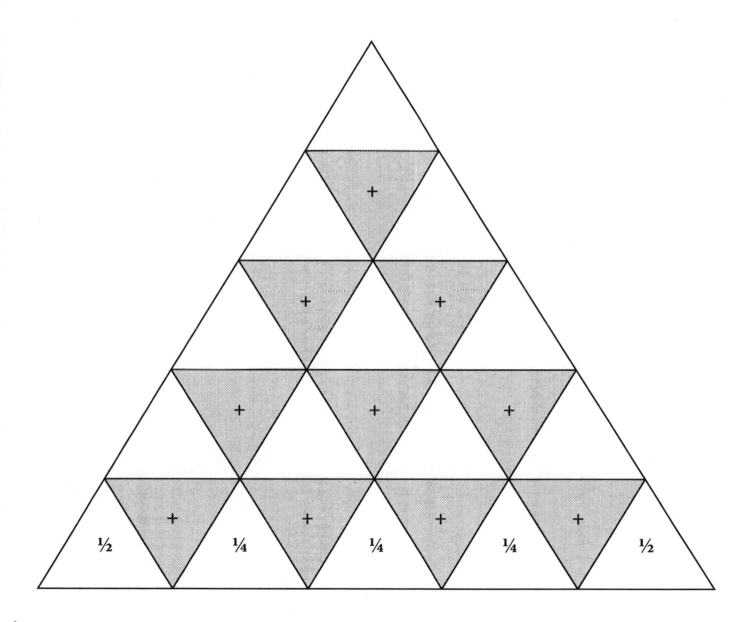

Not-So-Simple Fractions

 Add the fractions in each pair of small triangles to find the final sum. Keep all fractions in simplest form.

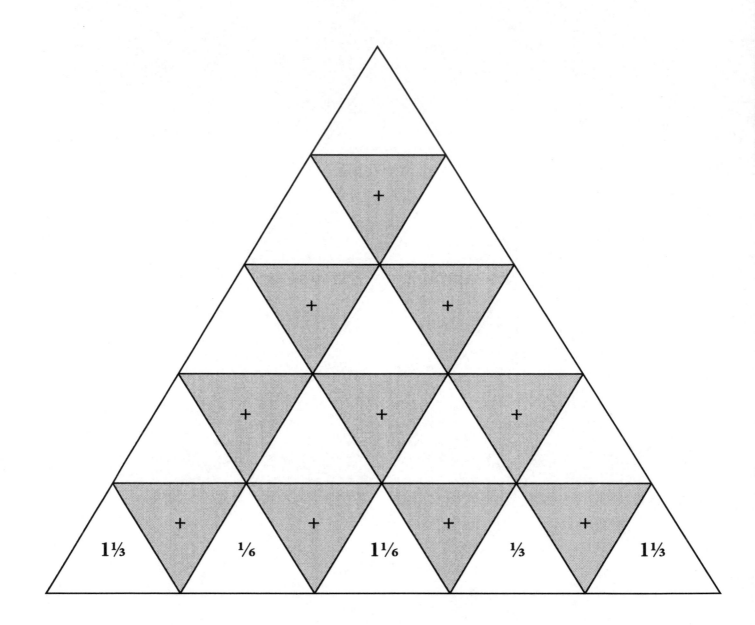

$1\frac{1}{3}$ $\frac{1}{6}$ $1\frac{1}{6}$ $\frac{1}{3}$ $1\frac{1}{3}$

Name _____   Date _____

Translation Association

Algebra is a mathematical language used to create models of real-world situations. Here is a representation of the number of pencils needed by a teacher for a test in which each student should have two pencils and the number of students is unknown: $2s$. The position of the number and the symbol mean "multiply." If the teacher wanted one pencil for each student plus two extra pencils, we would represent that as $s + 2$. And if we knew we had 60 pencils and wanted to know how many students the teacher could test if each student had two pencils, that would look like this: $60/2 = s$.

 **DIRECTIONS**   Match each description on the left with the algebraic symbol or expression it describes on the right.

1 _____ Sum of n and 3   **a)** $5b$

2 _____ quotient   **b)** $x - 9$

3 _____ product of 5 and b **c)** $n + 3$

4 _____ is   **d)** $2x$

5 _____ 9 less than x **e)** =

6 _____ twice   **f)** ÷

More Translation Association

English is read from left to right. It is not always correct to translate an algebraic equation into English in exactly the same order that it is written in the problem. While you *can* say that **x – 8** is "*x* take away 8," it is much more clear to say "8 less than *x*."

DIRECTIONS   Match each description on the left with the algebraic symbol or expression it describes on the right.

1 _____ a number increased by 2

a) $\frac{1}{2}n - 2$

2 _____ 8 less than *x*

b) $c \div {}^-3$

3 _____ three-fourths of a number

c) $x - 8$

4 _____ 2 less than half a number

d) $7 + f$

5 _____ a number decreased by 6

e) $n + 2$

6 _____ the quotient of *c* and ⁻3

f) $\frac{1}{2}(n + 1)$

7 _____ 3 more than twice a number

g) $\frac{3}{4}n$

8 _____ 7 and *f*

h) $n - 6$

9 _____ half of the quantity of *n* and 1

i) $2n + 3$

Translation with Equations

How do you translate grouping symbols? You can indicate parentheses with phrases like **the sum of**, **the difference of**, or **the quantity of**.

DIRECTIONS Match each description on the left with the algebraic expression it describes on the right.

1 _____ 10 less than half a number is 7.

a) $3n + 2 = 15$

2 _____ 9 decreased by twice a number is 1.

b) $2n - 9 = 1$

3 _____ 3 times the sum of a number and 2 is 15.

c) $n - (6 \times 1) = 18$

4 _____ Half the difference of 10 less than a number is 7.

d) $\dfrac{n - 10}{2} = 7$

5 _____ A number minus the product of 6 and 1 is 18.

e) $3(n + 2) = 15$

6 _____ The difference of 6 times a number and 1 is 18.

f) $\dfrac{n}{2} - 10 = 7$

7 _____ The difference of twice a number and 2 is 15.

g) $9 - 2n = 1$

8 _____ The sum of 3 times a number and 2 is 15.

h) $2n - 2 = 15$

9 _____ Nine less than twice a number is 1.

i) $6n - 1 = 18$

Name _____ Date _____

Multiplication Map

 **DIRECTIONS** Start anywhere along an edge. Make a "trail" by multiplying together a set of integers. Each trail must end up at and equal the number in the final box (-60). There is more than one correct path. Find as many trails as you can.

3	6	-4	5
-5	-1	-2	-3
-2	3	2	-1

	-60		

Multiplication rocks!

Name _____ Date _____

Division Map

 Start anywhere along an edge. Make a "trail" by dividing the first integer by an adjoining factor. Each trail must end up at and equal the number in the final boxes (⁻2 or 3). There is more than one correct path. Find as many trails as you can.

24	12	⁻2	⁻1	
18	3	1	⁻9	3
2	⁻2	⁻1	⁻3	
	⁻2			

Monomial Multiplication Map

Start anywhere along an edge. Make a "trail" by multiplying together a set of factors. Each trail must end up at and equal the number in the final box (16*n*). There is more than one correct path. Find as many trails as you can.

3	1	*n*	8
4	2	3	*n*
2	*n*	1	2
1	4	2	1

16*n*

Monomial Division Map

DIRECTIONS Start anywhere along an edge. Make a "trail" by dividing the first integer by an adjoining factor. Each trail must end up at and equal the number in the final box (2). There is more than one correct path. Find as many trails as you can.

$36a$	$576a^2b$	$64ab$	$6b$	
$24b$	$3a$	1	$16a$	
$4b$	3	$8a$	b	8
$3b$	2	1	a	
	2			

Name _____   Date _____

Position Problems

Variables are symbols that represent unknown quantities. There are other mathematical symbols that represent quantities. The best known may be the symbol for *Pi*, π. Pi stands for a very long number that we usually round off to 3.14.

 **DIRECTIONS** Use the key to decode the symbols and then solve each equation.

1	2	3
4	5	6
7	8	9

1

2 □ − □ =

3

4

5

Name _____ Date _____

More Position Problems

 Use the key to decode the symbols and then solve each equation.

8	-1	6
-3	7	2
5	9	4

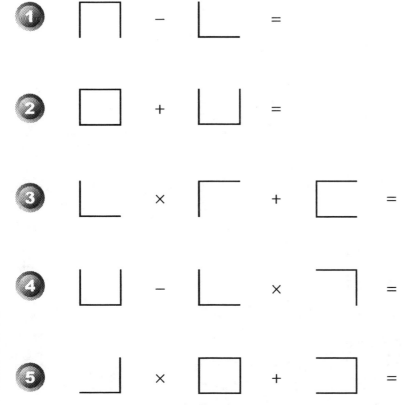

1 ⌐ – ⌐ =

2 □ + ⌐ =

3 ⌐ × ⌐ + ⌐ =

4 ⌐ – ⌐ × ⌐ =

5 ⌐ × □ + ⌐ =

Algebra Puzzles © 2006 Creative Teaching Press

Secret Code

 **DIRECTIONS** Use the key to decode the symbols and then solve each equation.

4	‑8	9
3	y	‑1
2	‑5	6

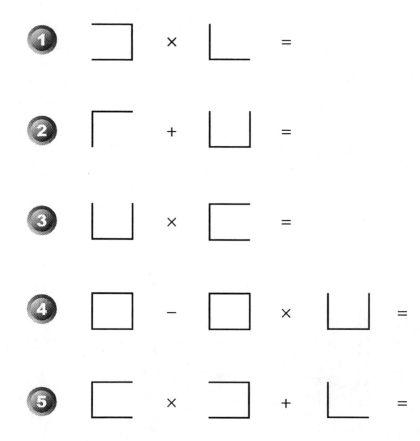

1 ⌐ × └ =

2 ⌐ + ⊔ =

3 ⊔ × ⌐ =

4 ☐ − ☐ × ⊔ =

5 ☐ × ⌐ + └ =

Name _____   Date _____

Code Name Math

 **DIRECTIONS**   Use the key to decode the symbols and then solve each equation.

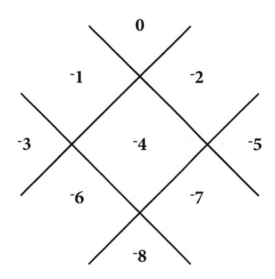

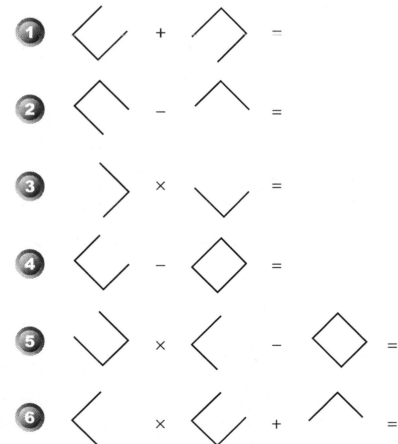

1   ⌞ + ⌝ =

2   ⌜ – ⌃ =

3   ⟩ × ⌄ =

4   ⌞ – ◇ =

5   ⌄⟩ × ⌜ – ◇ =

6   ⟨ × ⌞ + ⌃ =

Name _____   Date _____

Subtraction Mix-Up

 **DIRECTIONS** Subtract each pair of terms and record the difference in the table.

−	7	m	2	-3	8	1	6
4	-3						
1							
9							
n	n − 7						
3							
5							
2							

Name _____   Date _____

The Real Deal 1

Estimate: to approximate the value.

 **DIRECTIONS** The diagram represents an acre of real estate. The owner has sold the first fraction of his land for $7,000. Estimate the fraction of the whole that the remaining spaces represent and use that information to calculate a reasonable estimate of the value of the acre.

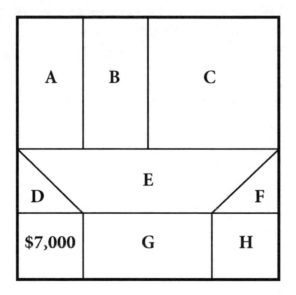

| A | B | C |

| E |
| D | | F |

| $7,000 | G | H |

1 A = _____

2 B = _____

3 C = _____

4 D = _____

5 E = _____

6 F = _____

7 G = _____

8 H = _____

9 Value of the whole: $_____

The Real Deal 2

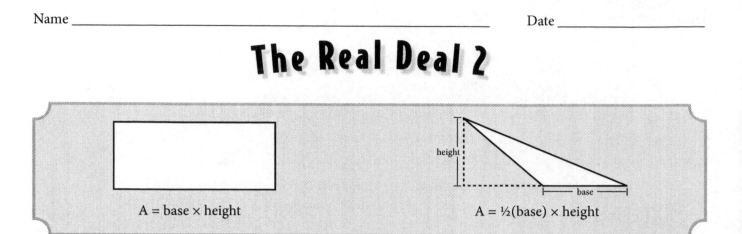

A = base × height   A = ½(base) × height

 The diagram represents an acre of real estate. The owner has sold the first fraction of her land for $x. Estimate the fraction of the whole that the remaining spaces represent and use that information to calculate a reasonable estimate of the value of the acre.

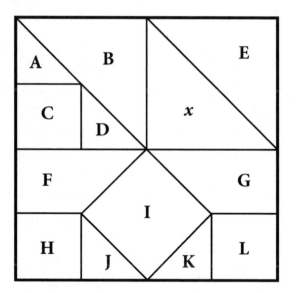

1 A = _____   **5** E = _____   **9** I = _____

2 B = _____   **6** F = _____   **10** J = _____

3 C = _____   **7** G = _____   **11** K = _____

4 D = _____   **8** H = _____   **12** L = _____

13 Value of the whole: $ _____

The Real Deal 3

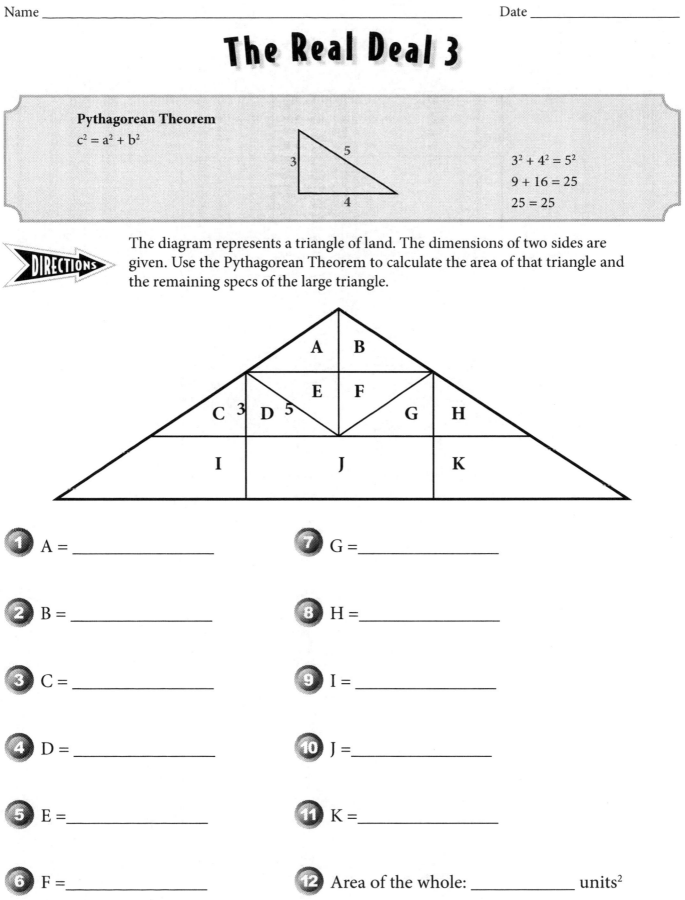

Pythagorean Theorem

$c^2 = a^2 + b^2$

$3^2 + 4^2 = 5^2$

$9 + 16 = 25$

$25 = 25$

DIRECTIONS The diagram represents a triangle of land. The dimensions of two sides are given. Use the Pythagorean Theorem to calculate the area of that triangle and the remaining specs of the large triangle.

1 A = _____

2 B = _____

3 C = _____

4 D = _____

5 E = _____

6 F = _____

7 G = _____

8 H = _____

9 I = _____

10 J = _____

11 K = _____

12 Area of the whole: _____ units²

Name _____ Date _____

The Real Deal 4

 **DIRECTIONS** Below are two boards. The dimensions of two sides of one triangle are given. Use the Pythagorean Theorem to calculate the total area of the shaded regions of each board.

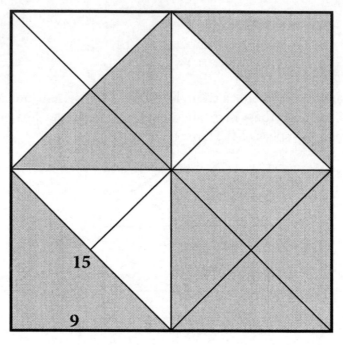

15

9

1 Total area of the shaded regions: _____ units^2

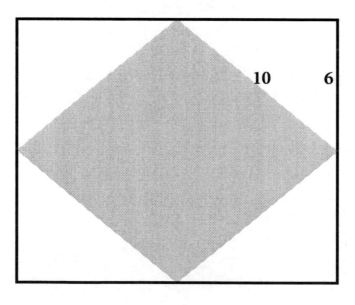

10   6

2 Total area of the shaded region: _____ units^2

Name _____ Date _____

The Real Deal 5

The radius of a circle is the distance from the center of the circle to its edge. The area of a circle is *pi* times the radius squared.

$$A = \pi r^2$$

 The diagram below represents a field of corn. Some joke players snuck into the field last night and drew BIG circles in the corn. The radius of the circles is 10 meters. Calculate the total area of the farmer's corn that was <u>not</u> disturbed. Use 3.14 for π.

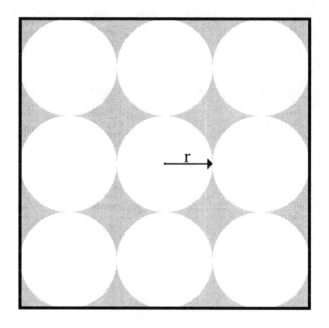

 1 The area of each circle is _____.

2 The length of each side of the square is _____.

3 The area of the square is _____.

4 The undisturbed portion of the farmer's land is _____.

Name _____ Date _____

Magic Square Mantra

In a magic square, the sum of each row, each column, and the two corner-to-corner diagonals is the same.

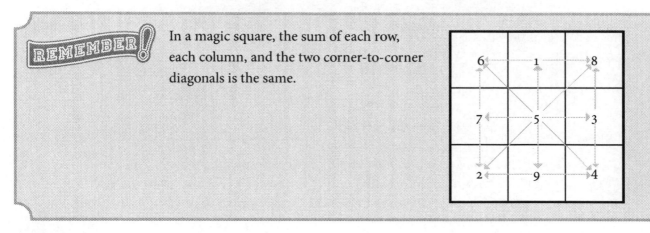

6	1	8
7	5	3
2	9	4

DIRECTIONS Simplify the expressions and equations. Then add any missing numbers to make the square "magic."

$2n = 8$ $n =$ _____		$3x - 9 = 18$ $x =$ _____	4^2
$2y + 5 = 35$ $y =$ _____			$\sqrt{9}$
$\sqrt{196}$		$a - 4 > 3$ $a >$ _____	
$\sqrt{1}$	$5 + b = 13$ $b =$ _____		$11 > 4r - 41$ $r <$ _____

Name _____ Date _____

Massive Magic Square

$\sqrt{25} = 5$ *and* $\sqrt{25} = -5$

 Simplify the expressions and equations. Then add any missing numbers to make the square "magic."

$9 \times 4 - 7$	$11 + 9^2 + 14$		$\sqrt{64}$
	$7 \times 7 - 6$	$^-6 + 7 \times 8$	
$12 + 8 \times 3$		$^-3 + 9 \times 9$	$^-3 + 8^2 - 4$
	$\sqrt{484}$		$9 \times 9 + 11$

Algebra Puzzles © 2006 Creative Teaching Press

Answer Key

Page 4

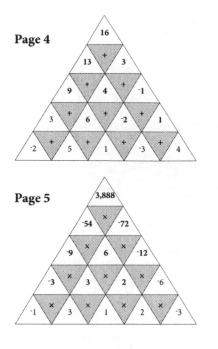

Page 5

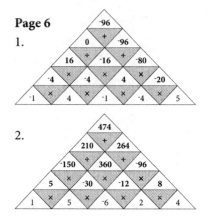

Page 6

1.

2.

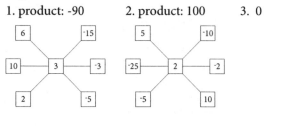

Page 7

1. product: -90   2. product: 100   3. 0

6		-15
10	3	-3
2		-5

5		-10
-25	2	-2
-5		10

Page 8

1.
2	4	6
8	4	0
2	4	6

2.
14	-7	8
-1	5	11
2	17	-4

3.
6	-6	6
2	2	2
-2	10	-2

4. 24

Possible responses:

5.
14	4	6
0	8	16
10	12	2

6.
7	9	-1
-3	5	13
11	1	3

7.
4	-9	5
1	0	-1
-5	9	-4

Page 9

1.
11	3	10
7	8	9
6	13	5

2.
4	9	11
15	8	1
5	7	12

3.
10	-3	11
7	6	5
1	15	2

4. yes

Page 10

×	5	-3	0	8	4	2	-9
9	45	-27	0	72	36	18	-81
1	5	-3	0	8	4	2	-9
6	30	-18	0	48	24	12	-54
5	25	-15	0	40	20	10	-45
-4	-20	12	0	-32	-16	-8	36
7	35	-21	0	56	28	14	-63
-3	-15	9	0	-24	-12	-6	27

Page 11

×	5	-1	n	12	-y	2	0
z	5z	-z	zn	12z	-zy	2z	0
-6	-30	6	-6n	-72	6y	-12	0
4	20	-4	4n	48	-4y	8	0
9	45	-9	9n	108	-9y	18	0
7	35	-7	7n	84	-7y	14	0
-3	-15	3	-3n	-36	3y	-6	0
x	5x	-1x	xn	12x	-xy	2x	0

Page 12

×	11	-3	8	-9	s	2	18
-7	-77	21	-56	63	-7s	-14	-126
1	11	-3	8	-9	s	2	18
-5	-55	15	-40	45	-5s	-10	-90
8	88	-24	64	-72	8s	16	144
-4	-44	12	-32	36	-4s	-8	-72
-s	-11s	3s	-8s	9s	-s²	-2s	-18s
15	165	-45	120	-135	15s	30	270

Page 13

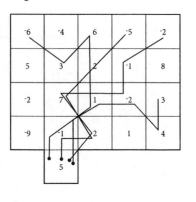

Page 14

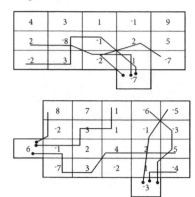

4	3	1	-1	9
2	-8	-1	2	5
-2	3	-2	1	-7

-7

8	7	1	-6	-5
-2	3	1	-1	-3
6	-1	2	4	2	5
-7	3	-2		-4

-3

Page 15

	$2x^2$	$3x$	$12x^2$	4	x
Positive Number				✓	
Quadratic Term	✓		✓		
Composite	✓	✓	✓	✓	
Factor of $24x^2$	✓	✓	✓	✓	✓
Reciprocal of ⅓x		✓			
Rational Number				✓	
A Constant				✓	

Page 16

	7	$a+2$	0	11	$2b^2$
Integer	✓		✓	✓	
Polynomial		✓			
Variable Expression		✓			✓
Equal to Its Absolute Value	✓		✓	✓	✓
Exponential Form					✓
Monomial	✓		✓	✓	✓
Prime	✓	✓		✓	

Page 17

1.   2.   3.   4.

Page 18

1. $4 - (^-3) - 6 = 1$
2. $^-11 + 3 - (^-7) = ^-1$
3. $^-4 \pm 0 - 14 = ^-18$
4. $23 + (^-21) + (^-12) = ^-10$
5. $8 - 7 - (^-5) = 6$
6. $^-21 - 14 + (^-3) = ^-38$

Page 19

1. $6 - 2 + 1 = 5$
2. $^-5 + (^-3) + n = n - 8$
3. $b - 9 - (^-8) = b - 17$
4. $7 + x - (^-2) = 9 + x$
5. $3 - 7 + (^-5) = ^-9$
6. $^-11 - 14 - (^-3) = ^-22$

Page 20

1. $^-4 - 5 + 8 = -1$
2. $2n + 2 - n = n + 2$
3. $^-n^2 + 5n + 2n^2 = 5n + n^2$
4. $2n^3 - n - n^3 = n^3 - n$
5. $3n^2 - 4n^2 + n^3 = n^3 - n^2$
6. $^-n - 3n - (^-3n^2) = 3n^2 - 4n$

Page 21

	A	B	C	D	E	F	G	H	I	J
1	1	4	8	$2n$	3	4	n	2	n	8
2	4	6	12	1	2	1	2	1	1	n
3	n	n	$6n$	1	4	6	n	2	n	3
4	4	1	3	$2n$	6	2	1	3	6	1
5	$3n$	4	n	1	2	$6n$	2	$3n$	2	$8n$
6	$2n$	4	2	12	3	n	12	1	$2n$	4

Page 22

	A	B	C	D	E	F	G	H	I	J
1	1	4	8	$2a$	3	4	a	2	^-a	8
2	4	$^-1$	12	1	2	1	$^-2$	$^-2$	1	a
3	a	$^-1$	$2a$	1	4	2	8	1	^-a	$^-2$
4	4	1	2	$^-8$	6	2	$^-1$	a	1	$^-1$
5	^-a	1	a	1	2	$4a$	2	$2a$	2	^-a
6	$2a$	$^-4$	2	$^-1$	1	a	^-a	1		$^-4$

Page 23

Answers will vary. Possible response:

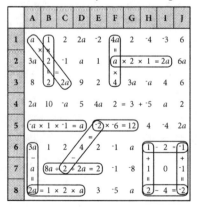

Page 24

Answers will vary. Possible response:

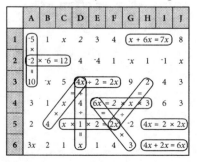

Page 25

Answers will vary. Check student work.

Page 26

Answers will vary. Check student work.

Page 27

Answers will vary. Check student work.

Page 28

Commutative: $1 + 2 = 2 + 1$; $7 \times 99 = 99 \times 7$

Associative: $2 + (4 + 6) = (2 + 4) + 6$; $(7 \times 5)3 = 7(5 \times 3)$

Distributive: $7(9 + 3) = 63 + 21$; $12(3 - 6) = 36 - 72$

Page 29

Identity Property of Multiplication: $12n \times 1 = 12n$

Identity Property of Addition:

$15{,}678{,}237 + (n \times 0) = 15{,}678{,}237$

Transitive: If $x + 4 = 5$ and $5 = 1 + 4$, then $x + 4 = 1 + 4$

Commutative: $43 \times 3 = 3 \times 43$

Associative: $(3 + n) + 2 = 3 + (n + 2)$

Distributive: $5(n - 3) = 5n - 15$

Page 30

+	n	4	2	-3	8	6	-1
7	n+7	11	9	4	15	13	6
-2	n-2	2	0	-5	6	4	-3
9	n+9	13	11	6	17	15	8
1	n+1	5	3	-2	9	7	0
3	n+3	7	5	0	11	9	2
m	n+m	m+4	m+2	m-3	m+8	m+6	m-1
-5	n-5	-1	-3	-8	3	1	-6

Page 31

+	a	1	2	13	-6	5	-8
12	12+a	13	14	25	6	17	4
-9	a-9	-8	-7	4	-15	-4	-17
-1	a-1	0	1	12	-7	4	-9
14	a+14	15	16	27	8	19	6
3	3+a	4	5	16	-3	8	-5
7	7+a	8	9	20	1	12	-1
-b	a-b	1-b	2-b	13-b	-6-b	5-b	-8-b

Page 32

+	-4	12	2	15	y	6	-1
-3	-7	9	-1	12	y-3	3	-4
-7	-11	5	-5	8	y-7	-1	-8
8	4	20	10	23	y+8	14	7
z	z-4	z+12	z+2	z+15	z+y	z+6	z-1
13	9	25	15	28	y+13	19	12
9	5	21	11	24	y+9	15	8
11	7	23	13	26	y+11	17	10

Page 33

1. Each cube is 2 lbs; $5n = 10$
2. Each cube is 5 lbs; $3n + 2 = 17$
3. Each stone is 0 lbs; $n = 0$

Page 34

1. The balance will tilt toward the blocks and cylinder.
2. The balance will tilt toward the block and cylinder.

Explanation:

First balance— $3x = 4y$; so $\frac{3}{4}x = y$.

Second balance— $x + 2y = z$; so $x + (\frac{3}{4}x + \frac{3}{4}x) = z$;

so $2\frac{1}{2}x = z$.

Third balance— $3x$ [?] $2y + z$; so $3x$ [?] $(\frac{3}{4}x + \frac{3}{4}x) + (2\frac{1}{2}x)$;

so $3x$ [<] $4x$.

Fourth balance— $3x$ [?] $y + z$; so $3x$ [?] $(\frac{3}{4}x) + (2\frac{1}{2}x)$;

so $3x$ [<] $3\frac{1}{4}x$.

Page 35

1. Possible response: If $3a = 2b$ and $a = 4c$, then $3(4c) = 2b$ so $12c/2 = 2b/2$; therefore, $6c = b$.
2. Answers will vary.

Page 36

1. $12 \times 3 \div 6 = 6$
2. $6 \times 2 \div (^-3) = ^-4$
3. $^-5 \times (^-3) \times 2 = 30$
4. $24 \div (^-8) \times 7 = ^-21$
5. $36 \div 9 \times (^-5) = ^-20$
6. $^-21 \div 7 \times (^-7) = 21$

Page 37

1. $^-7 \times 4 \div 2 = ^-14$
2. $10 \div (^-2) \times (^-8) = 40$
3. $15 \times 2 \times (^-2) = ^-60$
4. $4 \times (^-8) \div 8 = ^-4$
5. $12 \div 2 \div (^-6) = ^-1$
6. $^-9 \div (^-3) \times (^-3) = ^-9$

Page 38

1. $16 \div (^-4) \div (^-2) = 2$
2. $8 \times 5 \times (^-6) = ^-240$
3. $^-2 \div (^-2) \times 2 = 2$
4. $12 \div (^-3) \times 6 = ^-24$
5. $20 \div 2 \times (^-10) = ^-100$
6. $^-42 \div 7 \times (^-3) = 18$

Page 39

1. $2n^2 \times 2 \times n^2 = 4n^4$
2. $2a^3 \times 5b \times (^-b^3) = ^-10a^3b^4$
3. $^-n^2 \times (^-n) \times (^-n^3) = ^-n^6$
4. $8x^4 \div 2x \div (^-x^3) = ^-4$
5. $20n \div n^3 \times 10n^2 = 2n^2$
6. $30a^{10} \div 6a^4 \div a^5 = 5a$

Page 40

8

Page 41

103

Page 42

20

Page 43

2

Page 44

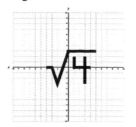

Page 45

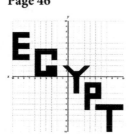

Page 46

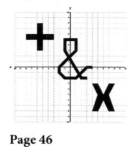

Page 47

Answers will vary. Check student work.

Page 48

Answers will vary. Check student work.

Page 49

Answers will vary. Check student work.

Page 50

Answers will vary. Check student work.

Page 51

1. a   2. e   3. i   4. p   5. s   *praise*

Page 52

1. a   2. g   3. l   4. o   5. s   *goals*

Page 53

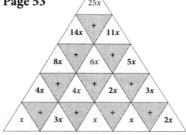

Page 54

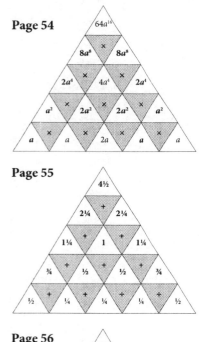

Page 55

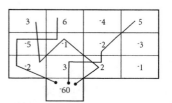

Page 56

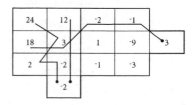

Page 57

1. c   2. f   3. a   4. e   5. b   6. d

Page 58

1. e   3. g   5. h   7. i   9. f
2. c   4. a   6. b   8. d

Page 59

1. f   3. e   5. c   7. h   9. b
2. g   4. d   6. i   8. a

Page 60

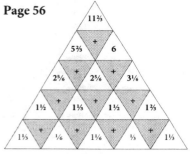

Page 61

Page 62

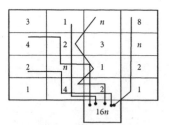

Page 63

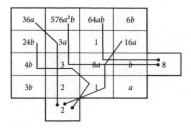

Page 64

1. $5 + 3 = 8$ 3. $2 \times 6 = 12$ 5. $3 \times 8 + 6 = 30$

2. $9 - 2 = 7$ 4. $3 - 3 = 0$

Page 65

1. $9 - 6 = 3$ 3. $6 \times 4 + 2 = 26$ 5. $8 \times 7 + (^-3) = 53$

2. $7 + (^-1) = 6$ 4. $^-1 - 6 \times 5 = ^-31$

Page 66

1. $3 \times 9 = 27$ 3. $^-8 \times (^-1) = 8$ 5. $^-1 \times 3 + 9 = 6$

2. $6 + (^-8) = ^-2$ 4. $y - y \times (^-8) = 9y$

Page 67

1. $^-2 + (^-6) = ^-8$ 3. $^-3 \times 0 = 0$ 5. $^-1 \times (^-5) - (^-4) = 9$

2. $^-7 - (^-8) = 1$ 4. $^-2 - (^-4) = 2$ 6. $^-5 \times (^-2) + (^-8) = 2$

Page 68

$-$	7	m	2	$^-3$	8	1	6
4	$^-3$	$4 - m$	2	7	$^-4$	3	$^-2$
1	$^-6$	$1 - m$	$^-1$	4	$^-7$	0	$^-5$
9	2	$9 - m$	7	12	1	8	3
n	$n-7$	$n-m$	$n-2$	$n+3$	$n-8$	$n-1$	$n-6$
3	$^-4$	$3 - m$	1	6	$^-5$	2	$^-3$
5	$^-2$	$5 - m$	3	8	$^-3$	4	$^-1$
2	$^-5$	$2 - m$	0	5	$^-6$	1	$^-4$

Page 69

1. A = $14,000   4. D = $3,500   7. G = $14,000

2. B = $14,000   5. E = $21,000   8. H = $7,000

3. C = $28,000   6. F = $3,500   9. Value: $112,000

Page 70

1. A = ¼x 6. F = ¾x 10. J = ¼x

2. B = x 7. G = ¾x 11. K = ¼x

3. C = ½x 8. H = ½x 12. L = ½x

4. D = ¼x 9. I = x 13. Value: 8x$

5. E = x

Page 71

1. A = 6   5. E = 6   9. I = 18

2. B = 6   6. F = 6   10. J = 24

3. C = 6   7. G = 6   11. K = 18

4. D = 6   8. H = 6   12. Area: 108 units^2

Page 72

1. 270 units^2 2. 96 units^2

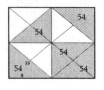 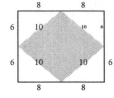

Page 73

1. 314 m   3. 3,600 m^2

2. 60 m   4. 774 m^2

Page 74

$2n=8$ $n=4$	5	$3x-9=18$ $x=9$	4^2 16
$2y+5=35$ $y=15$	10	6	$\frac{\sqrt{9}}{3}$
$\frac{\sqrt{196}}{14}$	11	$a-4>3$ $a>7$	2
$\frac{\sqrt{1}}{1}$	$3x-9=18$ $b=8$	12	$11>4r-41$ $r<13$

Page 75

$9 \times 4 - 7$ 29	$11 + 9^2 + 14$ 106	99	$\frac{\sqrt{64}}{8}$
64	$7 \times 7 - 6$ 43	$^-6 + 7 \times 8$ 50	85
$12 + 8 \times 3$ 36	71	$^-3 + 9 \times 9$ 78	$^-3 + 8^2 - 4$ 57
113	$\frac{\sqrt{484}}{22}$	15	$9 \times 9 + 11$ 92